COMMERCIAL PILOT
Practical Test Standards
for
ROTORCRAFT

• *HELICOPTER*

• *GYROPLANE*

FLIGHT STANDARDS SERVICE
Washington, DC 20591

NOTE

Material in FAA-S-8081-16A will be effective August 1, 2006. All previous editions of the Commercial Pilot—Rotorcraft (Helicopter and Gyroplane) Practical Test Standards will be obsolete as of this date.

FOREWORD

The Commercial Pilot—Rotorcraft (Helicopter and Gyroplane) Practical Test Standards (PTS) book has been published by the Federal Aviation Administration (FAA) to establish the standards for commercial pilot certification practical tests for the rotorcraft category, helicopter and gyroplane classes. FAA inspectors and designated pilot examiners shall conduct practical tests in compliance with these standards. Flight instructors and applicants should find these standards helpful during training and when preparing for the practical test.

Joseph K. intera, Manager

Regulatory Support Division

Flight Standards Service

CONTENTS

INTRODUCTION

General Information

The Flight Standards Service of the Federal Aviation Administration (FAA) has developed this practical test standard (PTS) to be used by examiners[1] when conducting commercial pilot—rotorcraft practical tests. Instructors are expected to use this PTS when preparing applicants for practical tests. Applicants should be familiar with this PTS and refer to these standards during their training.

The FAA gratefully acknowledges the valuable assistance provided by many individuals, companies, and organizations throughout the aviation community who have contributed their time and talent in assisting with the revision of this practical test standard.

Practical Test Standard Concept

Title 14 of the Code of Federal Regulations (14 CFR) part 61 species the areas in which knowledge and skill must be demonstrated by the applicant before the issuance of a commercial pilot certificate or rating. The CFRs provide the flexibility that permits the FAA to publish practical test standards containing the AREAS OF OPERATION and specific TASKs in which competency must be demonstrated. The FAA will revise this PTS whenever it is determined that changes are needed in the interest of safety. ***Adherence to the provisions of the regulations and the practical test standards is mandatory for the evaluation of commercial pilot applicants.***

Practical Test Book Description

This test book contains the following Commercial Pilot Practical Test Standards:

> **Section 1** Rotorcraft—Helicopter
> **Section 2** Rotorcraft—Gyroplane

The Commercial Pilot—Rotorcraft Practical Test Standards include the AREAS OF OPERATION and TASKs for the issuance of an initial commercial pilot certificate and for the addition of category and/or class ratings to that certificate.

AREAS OF OPERATION are phases of the practical test arranged in a logical sequence within this standard. They begin with Preflight Preparation and end with Post-flight Procedures. The examiner may conduct the practical test in any sequence that will result in a complete and efficient test. ***However, the ground portion of the practical test must be accomplished before the flight portion***.

TASKs are titles of knowledge areas, flight procedures, or maneuvers appropriate to an AREA OF OPERATION.

REFERENCE identifies the publication(s) that describe(s) the TASK. Descriptions of TASKs are not included in the standards because this information can be found in the current issue of the listed reference. Publications other than those listed may be used, for references if their content conveys substantially the same meaning as the referenced publications.

NOTE is used to emphasize special considerations required in the AREA OF OPERATION or TASK.

These practical test standards are based on the following reference

14 CFR part 43	Maintenance, Preventive Maintenance, Rebuilding, and Alteration
14 CFR part 61	Certification: Pilots and Flight Instructors
14 CFR part 67	Medical Standards and Certification
14 CFR part 91	General Operating and Flight Rules
NTSB part 830	Notification and Reporting of Aircraft Accidents and Incidents
FAA-H-8083-1	Aircraft Weight And Balance Handbook
FAA-H-8083-21	Rotorcraft Flying Handbook
FAA-H-8083-25	Pilot's Handbook of Aeronautical Knowledge
AC 00-6	Aviation Weather
AC 00-45	Aviation Weather Services
AC 60-22	Aeronautical Decision Making
AC 60-28	English Language Skill Standards Required by 14 CFR parts 61, 63, and 65
AC 61-65	Certification: Pilots and Flight Instructors and Ground Instructors
AC 61-84	Role of Preflight Preparation
AC 61-134	General Aviation Controlled Flight into Terrain Awareness
AC 90-48	Pilots' Role in Collision Avoidance
AC 90-87	Helicopter Dynamic Rollover
AC 90-95	Unanticipated right yaw in helicopters
AC 91-13	Cold Weather Operation of Aircraft
AC 91-32	Safety In and Around Helicopters
AC 91-42	Hazards of Rotating Propeller and Helicopter Rotor Blades
AC 91-55	Reduction of Electrical System failures following aircraft engine starting

AIM	Aeronautical Information Manual
AFD	Airport Facility Directory
FDC NOTAMs	National Flight data Center Notices to Airmen
Other	Pertinent Pilot's Operating Handbooks FAA-Approved Flight Manuals Navigation Charts

The Objective lists the important elements that must be satisfactorily performed to demonstrate competency in a TASK. The Objective includes:

1. specifically what the applicant should be able to do;
2. the conditions under which the TASK is to be performed; and
3. the acceptable standards of performance.

Abbreviations

14 CFR	Title 14 of the Code of Federal Regulations
ADM	Aeronautical Decision Making
AGL	Above Ground Level
ASOS	Automated Surface Observing System
ATC	Air Traffic Control
ATIS	Automatic Terminal Information Service
AWOS	Automated Weather Observing System
CFIT	Controlled Flight into Terrain
CG	Center of Gravity
CRM	Cockpit Resource Management
FA	Area Weather Forecast
FAA	Federal Aviation Administration
FDC	Flight Data Center
FSDO	Flight Standards District Office
GPO	Government Printing Office
ILS	Instrument Landing System
MEL	Minimum Equipment List
METAR	Aviation Routine Weather Report
NOTAM	Notice to Airmen
PTS	Practical Test Standard
RAIM	Receiver Autonomous Integrity Monitoring
RPM	Revolutions Per Minute

TAF	Terminal Aviation Forecast
VFR	Visual Flight Rules

Use Of The Practical Test Standards

The Commercial Pilot Rotorcraft Practical Test Standards have been designed to evaluate competency in both knowledge and skill. Commercial pilots are professionals engaged in various flight activities for compensation or hire. Because of their professional status, they should exhibit a significantly higher level of knowledge and skill than the private pilot. Although some TASKs listed are similar to those in the Private Pilot Rotorcraft Practical Test Standards, the wording used in the Commercial Pilot Rotorcraft Practical Test Standards reflects a higher level of competency expected of a commercial pilot applicant in performing these similar TASKs.

The FAA requires that all practical tests be conducted in accordance with the appropriate Commercial Pilot Practical Test Standards and the policies set forth in this INTRODUCTION. Commercial pilot applicants must be evaluated in **ALL** TASKs included in the AREAS OF OPERATION of the appropriate practical test standard unless otherwise noted.

An applicant, who holds at least a commercial pilot certificate seeking an additional rotorcraft category rating and/or class rating at the commercial pilot level will be evaluated in the AREAS OF OPERATION and TASKs listed in the Additional Rating Task Table. At the discretion of the examiner, an evaluation of the applicant's competence in the remaining AREAS OF OPERATION and TASKs may be conducted.

If the applicant holds two or more category or class ratings at least at the private level, and the Additional Rating Task Table indicates differing required TASKs, the "least restrictive" entry applies. For example, if "ALL" and "NONE" are indicated for one AREA OF OPERATION, the "NONE" entry applies. If "B" and "B, C" are indicated, the "B" entry applies.

In preparation for each practical test, the examiner must develop a written "plan of action." The "plan of action" is a tool, for the sole use of the examiner, to be used in evaluating the applicant. The plan of action need not be grammatically correct or in any formal format. The plan of action must contain all of the required AREAS OF OPERATION and TASKs and any optional TASKs selected by the examiner. The "plan of action" must incorporate one or more scenarios that will be used during the practical test.

The examiner should try to include as many of the TASKs into the scenario portion of the test as possible, but maintain the flexibility to change due to unexpected situations as they arise and still result in an efficient and valid test. ***Any TASK selected for evaluation during a practical test is to be evaluated in its entirety.***

The examiner is not required to follow the precise order in which the AREAS OF OPERATION and TASKs appear in this book. The examiner may change the sequence or combine TASKs with similar objectives to have an orderly and efficient flow of the practical test. For example, lost procedures may be combined with radio navigation. The examiner's "plan of action" should include the order and combination of TASKs to be demonstrated by the applicant in a manner that will result in an efficient and valid test.

The examiner is expected to use good judgment in the performance of simulated emergency procedures. The use of the safest means for simulation is expected. Consideration must be given to local conditions (both meteorological and topographical), at the time of the test, as well as the applicant's, workload, and the condition of the aircraft used. If the procedure being evaluated would jeopardize safety, it is expected that the applicant will simulate that portion of the maneuver.

Special Emphasis Areas

Examiners must place special emphasis upon areas of aircraft operation considered critical to flight safety. Among these are:

1. positive aircraft control;
2. procedures for positive exchange of flight controls (who is flying the aircraft);
3. collision avoidance;
4. wake turbulence avoidance;
5. runway incursion avoidance;
6. CFIT;
7. wire strike avoidance;
8. ADM and risk management;
9. checklist usage; and
10. other areas deemed appropriate to any phase of the practical test.

Although these areas may not be specifically addressed under each TASK, they are essential to flight safety and will be evaluated during the practical test. In all instances, the applicant's actions will relate to the complete situation.

Commercial Pilot Rotorcraft Practical Test Prerequisites

An applicant for the Commercial Pilot Rotorcraft Practical Test is required by 14 CFR part 61 to:

1. possess a private pilot certificate or meets the flight experience required for a private pilot certificate and pass the private helicopter or gyroplane knowledge and practical test;

2. have passed the appropriate commercial pilot knowledge test since the beginning of the 24th month before the month in which the practical test is completed;
3. obtain the applicable instruction and aeronautical experience prescribed for the commercial pilot certificate or rating sought;
4. possess at least a current third-class medical certificate issued under 14 CFR part 67;
5. be at least 18 years of age;
6. obtain a written statement from an appropriately certificated flight instructor certifying that the applicant has been given flight instruction in preparation for the practical test within 60 days preceding the date of application. The statement must also state that the instructor finds the applicant competent to pass the practical test and that the applicant has satisfactory knowledge of the subject area(s) in which a deficiency was indicated by the airman knowledge test report; and
7. be able to read, speak, write, and understand the English language. If there is a doubt, use AC 60-28, English Language Skill Standards.

Aircraft And Equipment Required for the Practical Test

The commercial pilot applicant is required by 14 CFR part 61, section 61.45 to provide an airworthy, certificated aircraft for use during the practical test. This section further requires that the aircraft must:

1. be of U.S., foreign or military registry of the same category, class, and type, if applicable, for the certificate and/or rating for which the applicant is applying;
2. have fully functioning dual controls, except as provided in 14 CFR part 61, section 61.45(c) and (e); in this CFR section; and
3. be capable of performing ALL AREAS OF OPERATION appropriate to the rating sought and have no operating limitations, which prohibit its use in any of the AREAS OF OPERATION, required for the practical test.

Flight Instructor Responsibility

An appropriately rated flight instructor is responsible for training the commercial pilot applicant to acceptable standards in **ALL** subject matter areas, procedures, and maneuvers included in the TASKs within the appropriate commercial pilot practical test standard.

Because of the impact of their teaching activities in developing safe, proficient pilots, flight instructors should exhibit a high level of knowledge, skill, and the ability to impart that knowledge and skill to students.

Throughout the applicant's training, the flight instructor is responsible for emphasizing the performance of effective visual scanning, collision avoidance, and runway incursion avoidance procedures.

Examiner Responsibility

The examiner conducting the practical test is responsible for determining that the applicant meets the acceptable standards of knowledge and skill of each TASK within the appropriate practical test standard. Since there is no formal division between the "oral" and "skill" portions of the practical test, this becomes an ongoing process throughout the test. To avoid unnecessary distractions, oral questioning, to determine the applicant's knowledge of TASKs and related safety factors, should be used judiciously at all times, especially during the flight portion of the practical test.

Examiners must test to the greatest extent practicable the applicant's correlative abilities rather than mere rote enumeration of facts throughout the practical test.

If the examiner determines that a TASK is incomplete, or the outcome uncertain, the examiner may require the applicant to repeat that TASK, or portions of that TASK. This provision has been made in the interest of fairness and does not mean that instruction, practice, or the repeating of an unsatisfactory TASK is permitted during the certification process.

Throughout the flight portion of the practical test, the examiner must evaluate the applicant's use of visual scanning and collision avoidance procedures.

Satisfactory Performance

Satisfactory performance to meet the requirements for certification is based on the applicant's ability to safely:

1. perform the TASKs specified in the AREAS OF OPERATION for the certificate or rating sought within the approved standards;
2. demonstrate mastery of the aircraft with the successful outcome of each TASK performed never seriously in doubt;
3. demonstrate satisfactory proficiency and competency within the approved standards;
4. demonstrate sound judgment and ADM; and
5. demonstrate single-pilot competence if the aircraft is type certificated for single-pilot operations.

Unsatisfactory Performance

The tolerances represent the performance expected in good flying conditions. If, in the judgment of the examiner, the applicant does not meet the standards of performance of any TASK performed, the associated AREA OF OPERATION is failed and therefore, the practical test is failed.

The examiner or applicant may discontinue the test at any time when the failure of an AREA OF OPERATION makes the applicant ineligible for the certificate or rating sought. ***The test may be continued ONLY with the consent of the applicant.*** If the test is discontinued, the applicant is entitled credit for only those AREAS OF OPERATION and their associated TASKs satisfactorily performed. However, during the retest and at the discretion of the examiner, any TASK may be re-evaluated including those previously passed.

Typical areas of unsatisfactory performance and grounds for disqualification are:

1. Any action or lack of action by the applicant that requires corrective intervention by the examiner to maintain safe flight.
2. Failure to use proper and effective visual scanning techniques to clear the area before and while performing maneuvers.
3. Consistently exceeding tolerances stated in the Objectives.
4. Failure to take prompt corrective action when tolerances are exceeded.

When a Notice of Disapproval is issued, the examiner will record the applicant's unsatisfactory performance and TASK(s) not completed in terms of AREA OF OPERATION(s) and specific TASK(s) not meeting the standard appropriate to the practical test conducted. The AREA(s) OF OPERATION/TASK(s) not tested and the number of practical test failures must also be recorded. If the applicant fails the practical test because of a special emphasis area, the Notice of Disapproval must indicate the associated TASK. i.e.: AREA OF OPERATION VIII, Settling-With-Power, failure to use proper collision avoidance procedures.

Letter of Discontinuance

When a practical test is discontinued for reasons other than unsatisfactory performance (i.e., equipment failure, weather, or illness) FAA Form 8700-1, Airman Certificate and/or Rating Application, and, if applicable, the Airman Knowledge Test Report, is to be returned to the applicant. The examiner at that time prepares, signs, and issues a Letter of Discontinuance to the applicant. The Letter of Discontinuance should identify the AREAS OF OPERATION and their associated TASKs of the practical test that were successfully completed. The applicant should be advised that the Letter of Discontinuance must be presented to the examiner when the practical test is resumed, and made part of the certification file.

Aeronautical Decision Making and Risk Management

Throughout the practical test, the examiner evaluates the applicant's ability to use good aeronautical decision-making procedures in order to identify risks. The examiner accomplishes this requirement by developing scenarios that incorporate as many TASKs as possible to evaluate the applicants risk management in making safe aeronautical decisions. For example, the examiner may develop a scenario that incorporates weather decisions and performance planning.

The applicant's ability to utilize all the assets available in making a risk analysis to determine the safest course of action is essential for satisfactory performance. The scenarios should be realistic and within the capabilities of the aircraft used for the practical test.

Single-Pilot Resource Management

Single-pilot resource management refers to the effective use of ALL available resources: human resources, hardware, and information. It is similar to crew resource management (CRM) procedures that are being emphasized in multi-crewmember operations except that only one crewmember (the pilot) is involved. Human resources "...includes all other groups routinely working with the pilot who are involved in decisions that are required to operate a flight safely. These groups include, but are not limited to: dispatchers, weather briefers, maintenance personnel, and air traffic controllers." Pilot resource management is not a single TASK; it is a set of skill competencies that must be evident in all TASKs in this practical test standard as applied to single-pilot operation.

Applicant's Use of Checklists

Throughout the practical test, the applicant is evaluated on the use of an appropriate checklist. Proper use is dependent on the specific TASK being evaluated. The situation may be such that the use of the checklist while accomplishing the elements of the Objective would be either unsafe or impractical, especially in a single-pilot operation. In this case, a review of the checklist after the elements have been accomplished would be appropriate. Division of attention and proper visual scanning would be considered when using a checklist.

Use Of Distractions During Practical Tests

Numerous studies indicate that many accidents have occurred when the pilot has been distracted during critical phases of flight. To evaluate the pilot's ability to utilize proper control technique while dividing attention both inside and/or outside the cockpit, the examiner should cause a realistic distraction during the **flight** portion of the practical test to evaluate the applicant's ability to divide attention while maintaining safe flight.

Positive Exchange of Flight Controls

During flight, there must always be a clear understanding between pilots of who has control of the aircraft. Prior to flight, a briefing should be conducted that includes the procedure for the exchange of flight controls. A positive three-step process in the exchange of flight controls between pilots is a proven procedure and one that is strongly recommended.

When one pilot wishes to give the other pilot control of the aircraft, he or she will say, “You have the flight controls.” The other pilot acknowledges immediately by saying, “I have the flight controls.” The first pilot again says, “You have the flight controls.” When control is returned to the first pilot, follow the same procedure. A visual check is recommended to verify that the exchange has occurred. There should never be any doubt as to who is flying the aircraft.

SECTION 1

COMMERCIAL PILOT ROTORCRAFT — HELICPTER

CONTENTS: SECTION 1

ADDITIONAL RATING TASK TABLE

Addition of a Rotorcraft/Helicopter rating to an existing Commercial Pilot Certificate								
Required TASKs are indicated by either the TASK letter(s) that apply(s) or an indication that all or none of the TASKs must be tested based on the notes in each AREA OF OPERATION.								
	COMMERCIAL PILOT RATING(S) HELD							
AREAS OF OPER-ATION	ASEL	ASES	AMEL	AMES	RG	Glider	Balloon	Airship
I	F,G,I,J	F,G,I,J	F,G,I,J	F,G,I,J	F,G,I,J	F,G,I,J	F,G,I,J	F,G,I,J
II	ALL	ALL	ALL	ALL	ALL	ALL	ALL	ALL
III	B,C	B,C	B,C	B,C	ALL	ALL	ALL	B,C
IV	ALL	ALL	ALL	ALL	ALL	ALL	ALL	ALL
V	ALL	ALL	ALL	ALL	ALL	ALL	ALL	ALL
VI	ALL	ALL	ALL	ALL	ALL	ALL	ALL	ALL
VII	NONE	NONE	NONE	NONE	B	B,C,D	B,C,D	NONE
VIII	ALL	ALL	ALL	ALL	ALL	ALL	ALL	ALL
IX	ALL	ALL	ALL	ALL	ALL	ALL	ALL	ALL
X	ALL	ALL	ALL	ALL	ALL	ALL	ALL	ALL

APPLICANT'S PRACTICAL TEST CHECKLIST (HELICOPTER) APPOINTMENT WITH EXAMINER:

EXAMINER'S NAME______________________________________

LOCATION ____

DATE/TIME ____

ACCEPTABLE AIRCRAFT

- □ Aircraft Documents:

  Airworthiness Certificate
  Registration Certificate
  Operating Limitations

- □ Aircraft Maintenance Records:

  Logbook Record of Airworthiness Inspections and AD Compliance

- □ Pilot's Operating Handbook and FAA-Approved

  Helicopter Flight Manual

- □ FCC Station License

PERSONAL EQUIPMENT

- □ View-Limiting Device
- □ Current Aeronautical Charts
- □ Computer and Plotter
- □ Flight Plan Form
- □ Flight Logs
- □ Current AIM, Airport Facility Directory, and Appropriate Publications

PERSONAL RECORDS

- □ Identification - Photo/Signature ID
- □ Pilot Certificate
- □ Current and Appropriate Medical Certificate

- □ Completed FAA Form 8710-1, Airman Certificate and/or Rating Application with Instructor's Signature (if applicable)
- □ AC Form 8080-2, Airman Written Test Report, or Computer Test Report
- □ Pilot Logbook with Appropriate Instructor Endorsements
- □ FAA Form 8060-5, Notice of Disapproval (if applicable)
- □ Approved School Graduation Certificate (if applicable)
- □ Examiner's Fee (if applicable)

EXAMINER'S PRACTICAL TEST CHECKLIST (HELICOPTER)

APPLICANT'S NAME____

LOCATION____

DATE/TIME____

I. PREFLIGHT PREPARATION

□ **A.** CERTIFICATES AND DOCUMENTS

□ **B.** AIRWORTHINESS REQUIREMENTS

□ **C.** WEATHER INFORMATION

□ **D.** CROSS-COUNTRY FLIGHT PLANNING

□ **E.** NATIONAL AIRSPACE SYSTEM

□ **F.** PERFORMANCE AND LIMITATIONS

□ **G.** OPERATION OF SYSTEMS

□ **H.** AEROMEDICAL FACTORS

□ **I.** PHYSIOLOGICAL ASPECTS OF NIGHT FLYING

□ **J.** LIGHTING AND EQUIPMENT FOR NIGHT FLYING

II. PREFLIGHT PROCEDURES

□ **A.** PREFLIGHT INSPECTION

□ **B.** COCKPIT MANAGEMENT

□ **C.** ENGINE STARTING AND ROTOR ENGAGEMENT

□ **D.** BEFORE TAKEOFF CHECK

III. AIRPORT AND HELIPORT OPERATIONS

□ **A.** RADIO COMMUNICATIONS AND ATC LIGHT SIGNALS

□ **B.** TRAFFIC PATTERNS

☐ **C.** AIRPORT/HELIPORT RUNWAY, HELIPORT, AND TAXIWAY SIGNS, MARKINGS, AND LIGHTING

IV. HOVERING MANEUVERS

☐ A. VERTICAL TAKEOFF AND LANDING

☐ B. SLOPE OPERATIONS

☐ C. SURFACE TAXI

☐ D. HOVER TAXI

☐ E. AIR TAXI

V. TAKEOFFS, LANDINGS, AND GO-AROUNDS

☐ A. NORMAL AND CROSSWIND TAKEOFF AND CLIMB

☐ B. NORMAL AND CROSSWIND APPROACH

☐ C. MAXIMUM PERFORMANCE TAKEOFF AND CLIMB

☐ D. STEEP APPROACH

☐ E. ROLLING TAKEOFF

☐ F. SHALLOW APPROACH AND RUNNING/ROLL-ON LANDING

☐ G. GO-AROUND

VI. PERFORMANCE MANEUVERS

☐ A. RAPID DECELERATION

☐ B. STRAIGHT IN ADUTOROTATION

☐ C. 180° AUTOROTATION

☐ D. APPROACH AND LANDING WITH SIMULATED POWERPLANT FAILURE – MULTIENGINE HELICOPTER

VII. NAVIGATION

☐ A. PILOTAGE AND DEAD RECKONING

☐ B. RADIO NAVIGATION AND RADAR SERVICES

- □ C. DIVERSION
- □ D. LOST PROCEDURES

VIII. EMERGENCY OPERATIONS

- □ A. POWER FAILURE AT A HOVER
- □ B. POWER FAILURE AT ALTITUDE
- □ C. SYSTEMS AND EQUIPMENT MALFUNCTIONS
- □ D. SETTLING-WITH-POWER
- □ E. LOW ROTOR RPM RECOVERY
- □ F. DYNAMIC ROLLOVER
- □ G. GROUND RESONANCE
- □ H. LOW G CONDITIONS
- □ I. EMERGENCY EQUIPMENT AND SURVIVAL GEAR

IX. SPECIAL OPERATIONS

- □ A. CONFINED AREA OPERATION
- □ B. PINNACLE/PLATFORM OPERATIONS

X. POST-FLIGHT PROCEDURES

- □ AFTER LANDING AND SECURING

I. AREA OF OPERATION: PREFLIGHT PREPARATION

A. TASK: CERTIFICATES AND DOCUMENTS

REFERENCE(S): 14 CFR parts 43, 61, 67, 91; FAA-H-8083-21, FAA-H-8083-25; POH/RFM.

Objective. To determine that the applicant exhibits knowledge of the elements related to certificates and documents by:

1. Explaining—
    a. commercial pilot certificate privileges, limitations, and recent flight experience requirements.
    b. medical certificate class and duration.
    c. pilot logbook or flight records.
2. Locating and explaining—
    a. airworthiness and registration certificates.
    b. operating limitations, placards, POH/RFM, and instrument markings.
    c. weight and balance data and equipment list.

B. TASK: AIRWORTHINESS REQUIREMENTS

REFERENCE(S): 14 CFR part 91; FAA-H-8083-21.

Objective. To determine that the applicant exhibits knowledge of the elements related to airworthiness requirements by:

1. Explaining—
    a. required instruments and equipment for day/night VFR.
    b. procedures and limitations for determining airworthiness of the helicopter with inoperative instruments and equipment with and without an MEL.
    c. requirements and procedures for obtaining a special flight permit.
2. Locating and explaining—
    a. airworthiness directives.
    b. compliance records.
    c. maintenance/inspection requirements.
    d. appropriate record keeping.

C. TASK: WEATHER INFORMATION

REFERENCE(S): AC 00-6, AC 00-45, AC 61-84; FAA-H-8083-25; AIM.

Objective. To determine that the applicant:

1. Exhibits knowledge of the elements related to weather information by analyzing available weather reports, charts, and forecasts from various sources with emphasis on—
    a. METAR, TAF, and FA.
    b. surface analysis chart.
    c. wind shear reports.
    d. winds and temperature aloft chart.
    e. AWOS, ASOS, and ATIS reports.
    f. significant weather prognostic charts.
2. Makes a competent "go/no-go" decision based on available weather information.

D. TASK: CROSS-COUNTRY FLIGHT PLANNING

REFERENCE(S): AC 61-84; FAA-H-8083-25; Navigation Charts; Airport/Facility Directory; FDC NOTAMs; AIM.

NOTE: In-flight demonstration of cross-country procedures by the applicant is tested under AREA OF OPERATION: NAVIGATION.

Objective. To determine that the applicant:

1. Exhibits knowledge of the elements related to cross-country flight planning by presenting and explaining a pre-planned VFR cross-country flight, as previously assigned by the examiner. On the day of the practical test, the flight plan should be to the first fuel stop necessary, based on maximum allowable passenger, baggage, and/or cargo loads using real-time weather.
2. Uses appropriate and current aeronautical charts.
3. Properly identifies airspace, obstructions, and terrain features, including discussion of wire strike avoidance techniques.
4. Selects easily identifiable en route checkpoints.
5. Selects most favorable altitudes, considering weather conditions and equipment capabilities.
6. Computes headings, flight time, and fuel requirements.
7. Selects appropriate navigation systems/facilities and communication frequencies.
8. Extracts and applies pertinent information from NOTAMs, Airport/Facility Directory, and other flight publications.
9. Completes a navigation log and simulates filing a VFR flight plan.

E. TASK: NATIONAL AIRSPACE SYSTEM

REFERENCE(S): 14 CFR parts 71, 91; Navigation Charts; AIM.

Objective. To determine that the applicant exhibits knowledge of the elements related to the national airspace system by explaining:

1. Basic VFR Weather Minimums – for all classes of airspace.
2. Airspace classes – their operating rules, pilot certification, and helicopter equipment requirements for the following—
    a. Class A.
    b. Class B.
    c. Class C.
    d. Class D.
    e. Class E.
    f. Class G.
3. Special use airspace and other airspace areas.

F. TASK: PERFORMANCE AND LIMITATIONS

REFERENCE(S): FAA-H-8083-1, FAA-H-8083-21; AC 91-23; POH/RFM.

Objective. To determine that the applicant:

1. Exhibits knowledge of the elements related to performance and limitations by explaining the use of charts, tables, and data to determine performance and the adverse effects of exceeding limitations.
2. Computes weight and balance. Determines the computed weight and center of gravity is within the helicopter's operating limitations and if the center of gravity will remain within limits during all phases of flight.
3. Demonstrates the use of appropriate performance charts, tables, and data.
4. Describes the effects of various atmospheric conditions on the helicopter's performance.
5. Understands the cause and effects of retreating blade stall.
6. Considers circumstances when operating within " avoid areas" of the height/velocity diagram.
7. Is aware of situations that lead to loss of tail rotor/antitorque effectiveness (unanticipated yaw).

G. TASK: OPERATION OF SYSTEMS

REFERENCE(S): FAA-H-8083-21; POH/AFM.

Objective. To determine that the applicant exhibits knowledge of the elements related to the appropriate normal operating procedures and limitations of the following systems by explaining:

1. Primary flight controls, trim, and, if installed, stability control.
2. Powerplant.
3. Main rotor and antitorque.
4. Landing gear, brakes, steering, skids, or floats, as applicable.
5. Fuel, oil, and hydraulic.
6. Electrical.
7. Pitot-static, vacuum/pressure and associated flight instruments, if applicable.
8. Environmental.
9. Anti-icing, including carburetor heat, if applicable.
10. Avionics equipment.

H. TASK: AEROMEDICAL FACTORS

REFERENCE(S): FAA-H-8083-25; AIM.

Objective. To determine that the applicant exhibits knowledge of the elements related to aeromedical factors by explaining:

1. The symptoms, causes, effects, and corrective actions of at least three (3) of the following—
    a.hypoxia.
    b.hyperventilation.
    c.middle ear and sinus problems.
    d.spatial disorientation.
    e.motion sickness.
    f.carbon monoxide poisoning.
    g.stress and fatigue.
    h.dehydration.
2. The effects of alcohol and drugs, including over-the- counter drugs.
3. The effects of nitrogen excesses during scuba dives upon a pilot and/or passenger in flight.

I. TASK: PHYSIOLOGICAL ASPECTS OF NIGHT FLYING

REFERENCE(S): FAA-H-8083-21, FAA-H-8083-25; AIM.

Objective. To determine that the applicant exhibits knowledge of the elements related to the physiological aspects of night flying by explaining:

1. The function of various parts of the eye essential for night vision.
2. Adaptation of the eye to changing light.
3. Correct use of the eye to accommodate changing light.
4. Coping with illusions created by various light conditions.
5. Effects of the pilot's physical condition on visual acuity.
6. Methods for increasing vision effectiveness.

J. TASK: LIGHTING AND EQUIPMENT FOR NIGHT FLYING

REFERENCE(S): FAA-H-8083-21, FAA-H-8083-25; POH/RFM.

Objective. To determine that the applicant:

1. Exhibits knowledge of the elements related to lighting and equipment for night flying by explaining—
   a. the types and uses of various personal lighting devices.
   b. the required equipment, and location of external navigation lighting of the helicopter.
   c. the meaning of various airport, heliport, and navigation lights, the method of determining their status, and the procedure for airborne activation of runway lights.
2. Locates and identifies switches, spare fuses, and circuit breakers pertinent to night operations.

II. AREA OF OPERATION: PREFLIGHT PROCEDURES

A. TASK: PREFLIGHT INSPECTION

REFERENCE(S): FAA-H-8083-21; POH/RFM.

Objective. To determine that the applicant:

1. Exhibits knowledge of the elements related to a preflight inspection. Including, which items must be inspected, the reasons for checking each item, and how to detect possible defects.
2. Inspects the helicopter with reference to an appropriate checklist.
3. Verifies that the helicopter is in condition for safe flight.

B. TASK: COCKPIT MANAGEMENT

REFERENCE(S): 14 CFR part 91; AC 91-32; POH/RFM.

Objective. To determine that the applicant:

1. Exhibits knowledge of the elements related to cockpit management procedures.
2. Ensures all loose items in the cockpit and cabin are secured.
3. Organizes material and equipment in an efficient manner so they are readily available.
4. Briefs the occupants on the use of safety belts, shoulder harnesses, doors, rotor blade avoidance, and emergency procedures.

C. TASK: ENGINE STARTING AND ROTOR ENGAGEMENT

REFERENCE(S): FAA-H-8083-21; AC 91-13, AC 91-42, AC 91-55; POH/RFM.

Objective. To determine that the applicant:

1. Exhibits knowledge of the elements related to correct engine starting procedures. Including, the use of an external power source, starting under various atmospheric conditions, awareness of other persons and property during start, and the effects of using incorrect starting procedures.
2. Ensures proper rotor blade clearance, and frictions flight controls, as necessary.
3. Utilizes the appropriate checklist for starting procedures.

D. TASK: BEFORE TAKEOFF CHECK

REFERENCE(S): FAA-H-8083-21; POH/RFM.

Objective. To determine that the applicant:

1. Exhibits knowledge of the elements related to the before takeoff check. Including, the reasons for checking each item and how to detect malfunctions.
2. Positions the helicopter properly considering other aircraft, wind, and surface conditions.
3. Divides attention inside and outside the cockpit.
4. Ensures that the engine temperature and pressure are suitable for run-up and takeoff.
5. Accomplishes the before takeoff check and ensures that the helicopter is in safe operating condition.
6. Reviews takeoff performance airspeeds, takeoff distances, departure, and emergency procedures.
7. Avoids runway incursions and/or ensures no conflict with traffic prior to takeoff.

III. AREA OF OPERATION: AIRPORT AND HELIPORT OPERATIONS

A. TASK: RADIO COMMUNICATIONS AND ATC LIGHT SIGNALS

REFERENCE(S): 14 CFR part 91; FAA-H-8083-25; AIM.

Objective. To determine that the applicant:

1. Exhibits knowledge of the elements related to radio communications and ATC light signals.
2. Selects appropriate frequencies.
3. Transmits using recommended phraseology.
4. Acknowledges radio communications and complies with instructions.

B. TASK: TRAFFIC PATTERNS

REFERENCE(S): 14 CFR part 91; FAA-H-8083-21; AIM, POH/RFM.

Objective. To determine that the applicant:

1. Exhibits knowledge of the elements related to traffic patterns. Including, procedures at airports and heliports with and without operating control towers, prevention of runway incursions collision avoidance, wake turbulence avoidance, and wind shear.
2. Complies with proper traffic pattern procedures.
3. Maintains proper spacing from other traffic or avoids the flow of fixed wing aircraft.
4. Corrects for wind drift to maintain proper ground track.
5. Maintains orientation with runway/landing area.
6. Maintains traffic pattern altitude ±100 feet, and appropriate airspeed, ±10 knots.

C. TASK: AIRPORT/HELIPORT RUNWAY, HELIPAD, AND TAXIWAY SIGNS, MARKINGS, AND LIGHTING

REFERENCE(S): 14 CFR part 91; FAA-H-8083-25; AIM.

Objective. To determine that the applicant:

1. Exhibits knowledge of the elements related to airport/heliport runway, and taxiway operations with emphasis on runway incursion avoidance.
2. Properly identifies and interprets airport/heliport, runway, and taxiway signs, markings, and lighting.

IV. AREA OF OPERATION: HOVERING MANEUVERS

A. TASK: VERTICAL TAKEOFF AND LANDING

REFERENCE(S): FAA-H-8083-21; AC 90-95; POH/RFM.

Objective. To determine that the applicant:

1. Exhibits knowledge of the elements related to a vertical takeoff to a hover and landing from a hover.
2. Ascends to and maintains recommended hovering altitude, and descends from recommended hovering altitude in headwind, crosswind, and tailwind conditions.
3. Maintains RPM within normal limits.
4. Establishes recommended hovering altitude, ±1/2 of that altitude within 10 feet of the surface; if above 10 feet, ±5 feet.
5. Avoids conditions that might lead to loss of tail rotor/antitorque effectiveness.
6. Keeps forward and sideward movement within 2 feet of a designated point, with no aft movement.
7. Descends vertically to within 2 feet of the designated touchdown point.
8. Maintains specified heading, ±10°.

B. TASK: SLOPE OPERATIONS

REFERENCE(S): FAA-H-8083-21; POH/RFM.

Objective. To determine that the applicant:

1. Exhibits knowledge of the elements related to slope operations.
2. Selects a suitable slope, approach, and direction considering wind effect, obstacles, dynamic rollover avoidance, and discharging passengers.
3. Properly moves toward the slope.
4. Maintains RPM within normal limits.
5. Makes a smooth positive descent to touch the upslope skid on the sloping surface.
6. Maintains positive control while lowering the downslope skid or landing gear to touchdown.
7. Recognizes when the slope is too steep and abandons the operation prior to reaching cyclic control stops.

8. Makes a smooth transition from the slope to a stabilized hover parallel to the slope.
9. Properly moves away from the slope.
10. Maintains the specified heading throughout the operation, ±5°.

C. TASK: SURFACE TAXI

REFERENCE(S): FAA-H-8083-21; AIM, POH/AFM.

NOTE: This TASK applies to only helicopters equipped with wheel- type landing gear.

Objective. To determine that the applicant:

1. Exhibits knowledge of the elements related to surface taxiing.
2. Surface taxies the helicopter from one point to another under headwind, crosswind, and tailwind conditions, with the landing gear in contact with the surface, avoiding conditions that might lead to loss of tail rotor/ antitorque effectiveness.
3. Properly uses cyclic, collective, and brakes to control speed while taxiing.
4. Properly positions nosewheel/tailwheel, if applicable, locked or unlocked.
5. Maintains RPM within normal limits.
6. Maintains appropriate speed for existing conditions.
7. Stops helicopter within ± 2 feet of a specified point.
8. Maintains specified track within ± 2 feet.

D. TASK: HOVER TAXI

REFERENCE(S): FAA-H-8083-21; AIM; POH/RFM.

Objective. To determine that the applicant:

1. Exhibits knowledge of the elements related to hover taxiing.
2. Hover taxies over specified ground references, demonstrating forward, sideward, and rearward hovering and hovering turns.
3. Maintains RPM within normal limits.
4. Maintains specified ground track within ± 2 feet on straight legs.
5. Maintains constant rate of turn at pivot points.
6. Maintains position within ± 2 feet of each pivot point during turns.
7. Makes 90°, 180°, and 360° pivoting turns, stopping within 10° of specified headings.
8. Maintains recommended hovering altitude, ±1/2 of that altitude within 10 feet of the surface, if above 10 feet, ±5 feet.

E. TASK: AIR TAXI

REFERENCE(S): FAA-H-8083-21; AC 90-95; AIM; POH/RFM.

Objective. To determine that the applicant:

1. Exhibits knowledge of the elements related to air taxiing.
2. Air taxies the helicopter from one point to another under headwind and crosswind conditions.
3. Maintains RPM within normal limits.
4. Selects a safe airspeed and altitude.
5. Maintains desired track and groundspeed in headwind and crosswind conditions, avoiding conditions that might lead to loss of tail rotor/antitorque effectiveness.
6. Maintains a specified altitude, ±5 feet.

V. AREA OF OPERATION: TAKEOFFS, LANDINGS, AND GO AROUNDS

A. TASK: NORMAL AND CROSSWIND TAKEOFF AND CLIMB

REFERENCE(S): FAA-H-8083-21; POH/RFM.

NOTE: If a calm wind weather condition exists, the applicant's knowledge of the crosswind elements must be evaluated through oral testing; otherwise a crosswind takeoff and climb must be demonstrated.

Objective. To determine that the applicant:

1. Exhibits knowledge of the elements related to normal and crosswind takeoff and climb, including factors affecting performance, to include height/velocity information.
2. Establishes a stationary position on the surface or a stabilized hover, prior to takeoff in headwind and crosswind conditions.
3. Maintains RPM within normal limits.
4. Accelerates to manufacturer's recommended climb airspeed, ±5 knots.
5. Maintains proper ground track with crosswind correction, as necessary.
6. Remains aware of the possibility of wind shear and/or wake turbulence.

B. TASK: NORMAL AND CROSSWIND APPROACH

REFERENCE(S): FAA-H-8083-21; POH/RFM.

NOTE: If a calm wind weather condition exists, the applicant's knowledge of the crosswind elements must be evaluated through oral testing; otherwise a crosswind approach and landing must be demonstrated.

Objective. To determine that the applicant:

1. Exhibits knowledge of the elements related to normal and crosswind approach.
2. Considers performance data, to include height/velocity information.
3. Considers the wind conditions, landing surface, and obstacles.
4. Selects a suitable termination point.
5. Establishes and maintains the normal approach angle, and rate of closure.
6. Remains aware of the possibility of wind shear and/or wake turbulence.
7. Avoids situations that may result in settling-with-power.
8. Maintains proper ground track with crosswind correction, as necessary.
9. Arrives at the termination point, on the surface or at a stabilized hover, ±2 feet.

C. TASK: MAXIMUM PERFORMANCE TAKEOFF AND CLIMB

REFERENCE(S): FAA-H-8083-21; POH/RFM.

Objective. To determine that the applicant:

1. Exhibits knowledge of the elements related to maximum performance takeoff and climb.
2. Considers situations where this maneuver is recommended and factors related to takeoff and climb performance, to include height/velocity information.
3. Maintains RPM within normal limits.
4. Utilizes proper control technique to initiate takeoff and forward climb airspeed attitude.
5. Utilizes the maximum available takeoff power.
6. After clearing all obstacles, transitions to normal climb attitude, airspeed, ±5 knots, and power setting.
7. Remains aware of the possibility of wind shear and/or wake turbulence.
8. Maintains proper ground track with crosswind correction, as necessary.

D. TASK: STEEP APPROACH

REFERENCE(S): FAA-H-8083-21; POH/RFM.

Objective. To determine that the applicant:

1. Exhibits knowledge of the elements related to a steep approach.
2. Considers situations where this maneuver is recommended and factors related to a steep approach, to include height/velocity information.
3. Considers the wind conditions, landing surface, and obstacles.
4. Selects a suitable termination point.
5. Establishes and maintains the recommended approach angle, (15° maximum) and proper rate of closure.
6. Avoids situations that can result in settling-with-power.
7. Remains aware of the possibility of wind shear and/or wake turbulence.
8. Maintains proper ground track with crosswind correction, if necessary.
9. Arrives at the termination point, on the surface or at a stabilized hover, ±2 feet.

E. TASK: ROLLING TAKEOFF

REFERENCE(S): FAA-H-8083-21; POH/RFM.

NOTE: This TASK applies only to helicopters equipped with wheel- type landing gear.

Objective. To determine that the applicant:

1. Exhibits knowledge of the elements related to a rolling takeoff.
2. Considers situations where this maneuver is recommended and factors related to takeoff and climb performance, to include height/velocity information.
3. Maintains RPM within normal limits.
4. Utilizes proper preparatory technique prior to initiating takeoff.
5. Initiates forward accelerating movement on the surface.
6. Transitions to a normal climb airspeed, ±5 knots, and power setting.
7. Remains aware of the possibility of wind shear and/or wake turbulence.
8. Maintains proper ground track with crosswind correction, if necessary.
9. Completes the prescribed checklist, if applicable.

F. TASK: SHALLOW APPROACH AND RUNNING/ROLL-ON LANDING

REFERENCE(S): FAA-H-8083-21; POH/RFM.

Objective. To determine that the applicant:

1. Exhibits knowledge of the elements related to shallow approach and running/roll-on landing, including the purpose of the maneuver, factors affecting performance data, to include height/velocity information, and effect of landing surface texture.
2. Maintains RPM within normal limits.
3. Considers obstacles and other hazards.
4. Establishes and maintains the recommended approach angle, and proper rate of closure.
5. Remains aware of the possibility of wind shear and/or wake turbulence.
6. Maintains proper ground track with crosswind correction, if necessary.
7. Maintains a speed that will take advantage of effective translational lift during surface contact with landing gear parallel with the ground track.
8. Utilizes proper flight control technique after surface contact.
9. Completes the prescribed checklist, if applicable.

G. TASK: GO-AROUND

REFERENCE(S): FAA-H-8083-21; POH/AFM.

Objective. To determine that the applicant:

1. Exhibits knowledge of the elements related to a go-around and when it is necessary.
2. Makes a timely decision to discontinue the approach to landing.
3. Maintains RPM within normal limits.
4. Establishes proper control input to stop descent and initiate climb.
5. Retracts the landing gear, if applicable, after a positive rate of climb indication.
6. Maintains proper ground track with crosswind correction, if necessary.
7. Transitions to a normal climb airspeed, ±5 knots.
8. Completes the prescribed checklist, if applicable.

VI. AREA OF OPERATION: PERFORMANCE MANEUVERS

NOTE: The examiner must select TASK A and at least one other TASK.

A. TASK: RAPID DECELERATION

REFERENCE(S): FAA-H-8083-21; Helicopter Flight Manual.

Objective. To determine that the applicant:

1. Exhibits knowledge of the elements related to rapid deceleration.
2. Maintains RPM within normal limits.
3. Properly coordinates all controls throughout the execution of the maneuver.
4. Maintains an altitude that will permit safe clearance between the tail boom and the surface.
5. Decelerates and terminates in a stationary hover at the recommended hovering altitude.
6. Maintains heading throughout the maneuver, ±5°.

B. TASK: STRAIGHT IN AUTOROTATION

REFERENCE(S): FAA-H-8083-21; POH/RFM.

Objective. To determine that the applicant:

1. Exhibits knowledge of the elements related to a straight in autorotation terminating with a power recovery to a hover.
2. Selects a suitable touchdown area.
3. Initiates the maneuver at the proper point.
4. Establishes proper aircraft trim and autorotation airspeed, ± 5 knots.
5. Maintains rotor RPM within normal limits.
6. Compensates for windspeed and direction as necessary to void undershooting or overshooting the selected landing area.
7. Utilizes proper deceleration, collective pitch application to a hover.
8. Comes to a hover within 100 feet of a designated point.

C. TASK: 180° AUTOROTATION

REFERENCE(S): FAA-H-8083-21; POH/RFM.

Objective. To determine that the applicant:

1. Exhibits knowledge of the elements related to a 180° autorotation terminating with a power recovery to a hover.
2. Selects a suitable touchdown area.
3. Initiates the maneuver at the proper point.
4. Establishes proper aircraft trim and autorotation airspeed, ±5 knots.
5. Maintains rotor RPM within normal limits.
6. Compensates for windspeed and direction as necessary to avoid undershooting or overshooting the selected landing area.
7. Utilizes proper deceleration, collective pitch application to a hover.
8. Comes to a hover within 100 feet of a designated point.

D. TASK: APPROACH AND LANDING WITH SIMULATED POWERPLANT FAILURE - MULTIENGINE HELICOPER

REFERENCE(S): FAA-H-8083-21; POH/RFM.

NOTE: In a multiengine helicopter maneuvering to a landing, the applicant should follow a procedure that simulates the loss of one powerplant.

Objective. To determine that the applicant:

1. Exhibits adequate knowledge of maneuvering to a landing with a powerplant inoperative, including the controllability factors associated with maneuvering, and the applicable emergency procedures.
2. Selects a suitable touchdown point.
3. Maintains, prior to beginning the final approach segment, the desired altitude ± 100 feet, the desired airspeed ± 10 knots, the desired heading ± 5° , and maintains desired track.
4. Establishes the approach and landing configuration appropriate for the runway or landing area, and adjusts the powerplant controls as required.
5. Maintains a normal approach angle and recommended airspeed to the point of transition to touchdown.
6. Terminates the approach in a smooth transition to touchdown.
7. Completes the after-landing checklist items in a timely manner, after clearing the landing area, and as recommended by the manufacturer.

VII. AREA OF OPERATION: NAVIGATION

A. TASK: PILOTAGE AND DEAD RECKONING

REFERENCE(S): FAA-H-8083-25; AC 61-84.

Objective. To determine that the applicant:

1. Exhibits knowledge of the elements related to pilotage and dead reckoning.
2. Follows the preplanned course by reference to landmarks.
3. Identifies landmarks by relating the surface features to chart symbols.
4. Navigates by means of precomputed headings, groundspeeds, and elapsed time.
5. Corrects for, and records, the differences between preflight fuel, groundspeed, and heading calculations and those determined en route.
6. Verifies the helicopter's position within three (3) nautical miles of the flight planned route.
7. Corrects for, and records, the differences between preflight fuel, groundspeed, and heading calculations and those determined en route.
8. Maintains the appropriate altitude, ±100 feet and established heading, ±10°.

B. TASK: RADIO NAVIGATION AND RADAR SERVICES

REFERENCE(S): FAA-H-8083-25; AC 61-84; Navigation Equipment Operation Manuals.

Objective. To determine that the applicant:

1. Exhibits knowledge of the elements related to radio navigation and ATC radar services.
2. Selects and identifies the appropriate facilities or coordinates, as appropriate.
3. Locates the helicopter's position relative to the navigation facilities or coordinates, as appropriate.
4. Intercepts and tracks a given radial or bearing.
5. Locates position using cross radials, coordinates, or bearings.
6. Recognizes and describes the indication of station or way point passage.
7. Recognizes signal loss and takes appropriate action.
8. Uses proper communication procedures when utilizing ATC radar services.
9. Maintains the appropriate altitude, ±100 feet.

C. TASK: DIVERSION

REFERENCE(S): FAA-H-8083-21; FAA-H-8083-25; AIM.

Objective. To determine that the applicant:

1. Exhibits knowledge of the elements related to procedures for diversion.
2. Selects an appropriate alternate airport or heliport and route.
3. Promptly, diverts toward the alternate airport or heliport.
4. Makes an accurate estimate of heading, groundspeed, arrival time, and fuel consumption to the alternate airport or heliport.
5. Maintains the appropriate altitude, ±100 feet and established heading, ±10°.

D. TASK: LOST PROCEDURES

REFERENCE(S): FAA-H-8083-21, FAA-H-8083-25; AC 61-84; AIM.

Objective. To determine that the applicant:

1. Exhibits knowledge of the elements related to lost procedures.
2. Selects an appropriate course of action.
3. Maintains an appropriate heading, and climbs, if necessary.
4. Attempts to identify prominent landmark(s).
5. Uses navigation systems/facilities and/or contacts an ATC facility for assistance as appropriate.
6. Plans a precautionary landing if deteriorating weather and/or fuel exhaustion is impending.

VIII. AREA OF OPERATION: EMERGENCY OPERATIONS

NOTE: Tasks F through I are knowledge only TASKs.

A. TASK: POWER FAILURE AT A HOVER

REFERENCE(S): FAA-H-8083-21; POH/RFM.

Objective. To determine that the applicant:

1. Exhibits knowledge of the elements related to power failure at a hover.
2. Determines that the terrain below the aircraft is suitable for a safe touchdown.
3. Performs autorotation from a stationary or forward hover into the wind at recommended altitude, and RPM, while maintaining established heading, ±5°.
4. Touches down with minimum sideward movement, and no rearward movement.
5. Exhibits orientation, division of attention, and proper planning.

B. TASK: POWER FAILURE AT ALTITUDE

NOTE: Simulated power failure at altitude must be given over areas where actual touchdowns can safely be completed in the event of an actual powerplant failure.

REFERENCE(S): FAA-H-8083-21; POH/RFM.

Objective. To determine that the applicant:

1. Exhibits knowledge of the elements related to power failure at altitude.
2. Establishes an autorotation and selects a suitable landing area.
3. Establishes proper aircraft trim and autorotation airspeed, ±5 knots.
4. Maintains rotor RPM within normal limits.
5. Compensates for windspeed and direction as necessary to avoid undershooting or overshooting the selected landing area.
6. Terminates approach with a power recovery at a safe altitude when directed by the examiner.

C. TASK: SYSTEMS AND EQUIPMENT MALFUNCTIONS

REFERENCE(S): FAA-H-8083-21; POH/RFM.

Objective. To determine that the applicant:

1. Exhibits knowledge of the elements related to causes, indications, and pilot actions for various systems and equipment malfunctions.
2. Analyzes the situation and takes action, appropriate to the helicopter used for the practical test, in at least four of the following areas—
   a. engine/oil and fuel.
   b. hydraulic, if applicable.
   c. electrical.
   d. carburetor or induction icing.
   e. smoke and/or fire.
   f. flight control/trim.
   g. pitot static/vacuum and associated flight instruments, if applicable.
   h. rotor and/or antitorque.
   i. various frequency vibrations and the possible components that may be affected.
   j. any other emergency unique to the helicopter flown.

D. TASK: SETTLING-WITH-POWER

REFERENCE(S): FAA-H-8083-21; POH/RFM.

Objective. To determine that the applicant:

1. Exhibits knowledge of the elements related to settling-with- power.
2. Selects an altitude that will allow recovery to be completed no less than 1,000 feet AGL or, if applicable, the manufacturer's recommended altitude, whichever is higher.
3. Promptly recognizes and announces the onset of settling with-power.
4. Utilizes the appropriate recovery procedure.

E. TASK: LOW ROTOR RPM RECOVERY

REFERENCE(S): FAA-H-8083-21; Appropriate Manufacturer's Safety Notices; POH/RFM.

NOTE: The examiner may test the applicant orally on this TASK if helicopter used for the practical test has a governor that cannot be disabled.

Objective. To determine that the applicant:

1. Exhibits knowledge of the elements related to low rotor RPM recovery, including the combination of conditions that are likely to lead to this situation.

2. Detects the development of low rotor RPM and initiates prompt corrective action.

3. Utilizes the appropriate recovery procedure.

F. TASK: DYNAMIC ROLLOVER

REFERENCE(S): FAA-H-8083-21; AC 90-87; POH/RFM.

Objective. To determine that the applicant:

1. Exhibits knowledge of the elements related to the aerodynamics of dynamic rollover.

2. Understands the interaction between the antitorque thrust, crosswind, slope, CG, cyclic and collective pitch control in contributing to dynamic rollover.

3. Explains preventive flight technique during takeoffs, landings, and slope operations.

G. TASK: GROUND RESONANCE

REFERENCE(S): FAA-H-8083-21; POH/RFM.

Objective. To determine that the applicant:

1. Exhibits knowledge of the elements related to a fully articulated rotor system and the aerodynamics of ground resonance.

2. Understands the conditions that contribute to ground resonance.

3. Explains preventive flight technique during takeoffs and landings.

H. TASK: LOW G CONDITIONS

REFERENCE(S): Helicopter Flight Manual.

Objective. To determine that the applicant:

1. Exhibits knowledge of the elements related to low G conditions.

2. Understands and recognizes the situations that contribute to low G conditions.

3. Explains proper recovery procedures.

I. TASK: EMERGENCY EQUIPMENT AND SURVIVAL GEAR

REFERENCE(S): FAA-H-8083-21; POH/RFM.

Objective. To determine that the applicant:

1. Exhibits knowledge of the elements related to emergency equipment and survival gear appropriate to the helicopter environment encountered during flight.
2. Identifies appropriate equipment that should be on board the helicopter.

IX. AREA OF OPERATION: SPECIAL OPERATIONS

A. TASK: CONFINED AREA OPERATION

REFERENCE(S): FAA-H-8083-21; POH/RFM.

Objective. To determine that the applicant:

1. Exhibits knowledge of the elements related to confined area operations.
2. Accomplishes a proper high and low reconnaissance.
3. Selects a suitable approach path, termination point, and departure path.
4. Tracks the selected approach path at an acceptable approach angle and rate of closure to the termination point.
5. Maintains RPM within normal limits.
6. Avoids situations that can result in settling-with-power.
7. Terminates at a hover or on the surface, as conditions allow.
8. Accomplishes a proper ground reconnaissance.
9. Selects a suitable takeoff point, considers factors affecting takeoff and climb performance under various conditions.

B. TASK: PINNACLE/PLATFORM OPERATIONS

REFERENCE(S): FAA-H-8083-21; POH/RFM.

Objective. To determine that the applicant:

1. Exhibits knowledge of the elements related to pinnacle/platform operations.
2. Accomplishes a proper high and low reconnaissance.
3. Selects a suitable approach path, termination point, and departure path.
4. Tracks the selected approach path at an acceptable approach angle and rate of closure to the termination point.
5. Maintains RPM within normal limits.
6. Terminates at a hover or on the surface, as conditions allow.
7. Accomplishes a proper ground reconnaissance.
8. Selects a suitable takeoff point, considers factors affecting takeoff and climb performance under various conditions.

X. AREA OF OPERATION: POST-FLIGHT PROCEDURES

A. TASK: AFTER LANDING AND SECURING

REFERENCE(S): FAA-H-8083-21; POH/RFM.

Objective. To determine that the applicant:

1. Exhibits knowledge of the elements related to after-landing, parking, and securing.
2. Minimizes the hazardous effects of rotor downwash during hovering.
3. Parks in an appropriate area, considering the safety of nearby persons and property.
4. Follows the appropriate procedure for engine shutdown.
5. Completes the appropriate checklist.
6. Conducts an appropriate postflight inspection and secures the aircraft.

SECTION 2

COMMERCIAL PILOT ROTORCRAFT — GYROPLANE

CONTENTS: SECTION 2

ADDITIONAL RATING TASK TABLE

Addition of an a Rotorcraft/Gyroplane rating to an existing Commercial Pilot Certificate								
Required TASKs are indicated by either the TASK letter(s) that apply(s) or an indication that all or none of the TASKs must be tested based on the notes in each AREA OF OPERATION.								
COMMERCIAL PILOT RATING(S) HELD								
AREAS OF OPER-ATION	**ASEL**	**ASES**	**AMEL**	**AMES**	**RH**	**Glider**	**Balloon**	**Airship**
I	A,F,G	A,F,G	A,F,G	A,F,G	A,F,G	A,F,G	A,F,G	A,F,G
II	ALL	ALL	ALL	ALL	ALL	ALL	ALL	ALL
III	B	B,C	B	B,C	B	B	ALL	B
IV	ALL	ALL	ALL	ALL	ALL	ALL	ALL	ALL
V	ALL	ALL	ALL	ALL	ALL	ALL	ALL	ALL
VI	ALL	ALL	ALL	ALL	ALL	ALL	ALL	ALL
VII	ALL	ALL	ALL	ALL	ALL	ALL	ALL	ALL
VIII	ALL	ALL	ALL	ALL	ALL	ALL	ALL	ALL
IX	ALL	ALL	ALL	ALL	ALL	ALL	ALL	ALL

APPLICANT'S PRACTICAL TEST CHECKLIST (GYROPLANE) APPOINTMENT WITH EXAMINER:

EXAMINER'S NAME____

LOCATION ____

DATE/TIME____

ACCEPTABLE AIRCRAT

- □ Aircraft Documents:
  - Airworthiness Certificate
  - Registration Certificate
  - Operating Limitations
- □ Aircraft Maintenance Records:
  - Logbook Record of Airworthiness Inspections and AD Compliance
- □ Pilot's Operating Handbook and FAA-Approved
  - Gyroplane Flight Manual
- □ FCC Station License

PERSONAL EQUIPMENT

- □ View-Limiting Device
- □ Current Aeronautical Charts
- □ Computer and Plotter Flight Plan Form
- □ Flight Logs
- □ Current AIM, Airport Facility Directory, and Appropriate Publications

PERSONAL RECORDS

- □ Identification - Photo/Signature ID
- □ Pilot Certificate
- □ Current and Appropriate Medical Certificate
- □ Completed FAA Form 8710-1, Airman Certificate and/or

☐ Rating Application with Instructor's Signature (if applicable)

☐ AC Form 8080-2, Airman Written Test Report, or Computer Test Report

☐ Pilot Logbook with Appropriate Instructor Endorsements

☐ FAA Form 8060-5, Notice of Disapproval (if applicable)

☐ Approved School Graduation Certificate (if applicable)

☐ Examiner's Fee (if applicable)

EXAMINER'S PRACTICAL TEST CHECKLIST (GYROPLANE)

APPLICANT'S NAME____

LOCATION____

DATE/TIME____

I. PREFLIGHT PREPARATION

- **A.** CERTIFICATES AND DOCUMENTS
- **B**. AIRWORTHINESS REQUIRMENTS
- **C.** WEATHER INFORMATION
- **D.** CROSS-COUNTRY FLIGHT PLANNING
- **E.** NATIONAL AIRSPACE SYSTEM
- **F.** PERFORMANCE AND LIMITATIONS
- **G.** OPERATION OF SYSTEMS
- **H.** AEROMEDICAL FACTORS
- **I.** PHYSIOLOGICAL ASPECTS OF NIGHT FLYING
- **J.** LIGHTING AND EQUIPMENT FOR NIGHT FLYING

II. PREFLIGHT PROCEDURES

- A. PREFLIGHT INSPECTION
- B. COCKPIT MANAGEMENT
- C. ENGINE STARTING
- D. TAXIING
- E. BEFORE TAKEOFF CHECK

III. AIRPORT OPERATIONS

- **A.** RADIO COMMUNICATIONS AND ATC LIGHT SIGNALS
- **B**. TRAFFIC PATTERNS
- **C**. AIRPORT MARKINGS AND LIGHTING

IV. TAKEOFFS, LANDINGS, AND GO-AROUNDS

- **A.** NORMAL AND CROSSWIND TAKEOFF AND CLIMB
- **B.** NORMAL AND CROSSWIND APPROACH AND LANDING
- **C.** SOFT-FIELD TAKEOFF AND CLIMB
- **D.** SOFT-FIELD APPROACH AND LANDING
- **E.** SHORT-FIELD TAKEOFF AND CLIMB
- **F.** SHORT-FIELD APPROACH AND LANDING
- **G.** GO-AROUND

V. PERFORMANCE MANEUVER

- **A.** STEEP TURNS

VI. NAVIGATION

A. PILOTAGE AND DEAD RECKONING
B. RADIO NAVIGATION AND RADAR SERVICES
C. DIVERSION
D. LOST PROCEDURES

VII. FLIGHT AT SLOW AIRSPEED

A. STRAIGHT-AND-LEVEL, TURNS, CLIMBS, AND DESCENTS AT SLOW AIRSPEEDS
B. HIGH RATE OF DESCENT AND RECOVERY

VIII. EMERGENCY OPERATIONS

A. EMERGENCY APPROACH AND LANDING
B. LIFT-OFF AT LOW AIRSPEED AND HIGH ANGLE OF ATTACK
C. GROUND RESONANCE
D. SYSTEMS AND EQUIPMENT MALFUNCTIONS
E. EMERGENCY EQUIPMENT AND SURVIVAL GEAR

IX. POST-FLIGHT PROCEDURES

A. AFTER LANDING, PARKING, AND SECURING

I. AREA OF OPERATION: PREFLIGHT PREPARATION

A. TASK: CERTIFICATES AND DOCUMENTS

REFERENCE(S): 14 CFR parts 43, 61, 67, 91; FAA-H-8083-21, FAA-H-8083-25; Gyroplane Flight Manual.

Objective. To determine that the applicant exhibits knowledge of the elements related to certificates and documents by:

1. Explaining—
   a. commercial pilot certificate privileges and limitations and recent flight experience requirements.
   b. medical certificate class and duration.
   c. pilot logbook or flight records.
2. Locating and explaining—
   a. airworthiness and registration certificates.
   b. operating limitations, placards, instrument markings, and gyroplane flight manual.
   c. weight and balance data and equipment list.

B. TASK: AIRWORTHINESS REQUIREMENTS

REFERENCE(S): 14 CFR part 91; FAA-H-8083-21.

Objective. To determine that the applicant exhibits knowledge of the elements related to airworthiness requirements by:

1. Explaining—
   a. required instruments and equipment for day/night VFR.
   b. procedures and limitations for determining airworthiness of the gyroplane with inoperative instruments and equipment with and without an MEL.
   c. requirements and procedures for obtaining a special flight permit.
2. Locating and explaining—
   a. airworthiness directives.
   b. compliance records.
   c. maintenance/inspection requirements.
   d. appropriate record keeping.

C. TASK: WEATHER INFORMATION

REFERENCE(S): 14 CFR 91; AC 00-6, AC 00-45, AC 61-84; FAA-H-8083-25; AIM.

Objective. To determine that the applicant:

1. Exhibits knowledge of the elements related to weather information by analyzing weather reports, charts, and forecasts from various sources with emphasis on—
   a. METAR, TAF, and FA.
   b. surface analysis chart.
   c. radar summary chart.
   d. winds and temperature aloft chart.
   e. significant weather prognostic charts.
   f. AWOS, ASOS, and ATIS reports.
2. Makes a competent " go/no-go" decision based on available weather information.

D. TASK: CROSS-COUNTRY FLIGHT PLANNING

REFERENCE(S): 14 CFR part 91; FAA-H-8083-25; AC 61-84; Navigation Charts; Airport/Facility Directory; FDC NOTAMs; AIM.

NOTE: In-flight demonstration of cross-country procedures by the applicant is tested under AREA OF OPERATION: NAVIGATION.

Objective. To determine that the applicant:

1. Exhibits knowledge of the elements related to cross-country flight planning by presenting and explaining a pre-planned VFR cross-country flight, as previously assigned by the examiner. On the day of the practical test, the final flight plan shall be to the first fuel stop necessary, based on maximum allowable passenger, baggage, and/or cargo loads using real time weather.
2. Uses appropriate and current aeronautical charts.
3. Properly identifies airspace, obstacles, and terrain features, including discussion of wire strike avoidance techniques.
4. Selects easily identifiable en route checkpoints.
5. Selects the most favorable altitudes, considering weather conditions and equipment capabilities.
6. Computes headings, flight time, and fuel requirements.
7. Selects appropriate navigation systems/facilities and communication frequencies.
8. Applies pertinent information from FDC NOTAMs, Airport/Facility Directory, and other flight publications.
9. Completes a navigation log and simulates filing a VFR flight plan.

E. TASK: NATIONAL AIRSPACE SYSTEM

REFERENCE(S): 14 CFR parts 71, 91; Navigation Charts; AIM.

Objective. To determine that the applicant exhibits knowledge of the elements related to the national airspace system by explaining:

1. Basic VFR Weather Minimums – for all classes of airspace.
2. Airspace classes – their boundaries, pilot certification, and gyroplane equipment requirements for the following—
    a. Class A.
    b. Class B.
    c. Class C.
    d. Class D.
    e. Class E.
    f. Class G.
3. Special use airspace and other airspace areas.

F. TASK: PERFORMANCE AND LIMITATIONS

REFERENCE(S): FAA-H-8083-1, FAA-H-8083-21; AC 61-84; Gyroplane Flight Manual.

Objective. To determine that the applicant:

1. Exhibits knowledge of the elements related to performance and limitations by explaining the use of charts, tables, and data to determine performance and the adverse effects of exceeding limitations.
2. Computes weight and balance. Determines the computed weight and center of gravity is within the gyroplane's operating limitations and if the weight and center of gravity will remain within limits during all phases of flight.
3. Demonstrates the use of appropriate performance charts, tables, and data.
4. Describes the effects of atmospheric conditions on the gyroplane's performance.
5. Understands the cause, effect, and avoidance procedure of " power pushover," and " pilot induced oscillation."

G. TASK: OPERATION OF SYSTEMS

REFERENCE(S): FAA-H-8083-25; Gyroplane Flight Manual.

Objective. To determine that the applicant exhibits knowledge of the elements related to the operation of systems on the gyroplane provided for the flight test by explaining at least four (4) of the following systems selected by the examiner.

1. Primary flight controls and trim.
2. Powerplant.
3. Rotor, including prerotator/spin-up control, if applicable.
4. Landing gear, brakes, and steering.
5. Fuel, oil, and hydraulic.
6. Electrical.
7. Pitot-static, vacuum/pressure, and associated flight instruments, if applicable.
8. Environmental, if applicable.
9. Anti-icing, including carburetor heat, if applicable.
10. Avionics equipment.

H. TASK: AEROMEDICAL FACTORS

REFERENCE(S): FAA-H-8083-25; AIM.

Objective. To determine that the applicant exhibits knowledge of the elements related to aeromedical factors by explaining:

1. The symptoms, causes, effects, and corrective actions of at least four (4) of the following—
    a. hypoxia.
    b. hyperventilation.
    c. middle ear and sinus problems.
    d. spatial disorientation.
    e. motion sickness.
    d. carbon monoxide poisoning.
    e. stress and fatigue.
2. The effects of alcohol and drugs, including over-the-counter drugs.
3. The effects of nitrogen excesses during scuba dives upon a pilot and/or passenger in flight.

I. TASK: PHYSIOLOGICAL ASPECTS OF NIGHT FLYING

REFERENCE(S): FAA-H-8083-21, FAA-H-8083-25; AIM.

Objective. To determine that the applicant exhibits knowledge of the elements related to the physiological aspects of night flying by explaining:

1. The function of various parts of the eye essential for night vision.
2. Adaptation of the eye to changing light.
3. Correct use of the eye to accommodate changing light.
4. Coping with illusions created by various light conditions.
5. Effects of the pilot's physical condition on visual acuity.
6. Methods for increasing vision effectiveness.

J. TASK: LIGHTING AND EQUIPMENT FOR NIGHT FLYING

REFERENCE(S): FAA-H-8083-21, FAA-H-8083-25; AIM; Gyroplane Flight Manual.

Objective. To determine that the applicant:

1. Exhibits knowledge of the elements related to lighting and equipment for night flying by explaining—
    a. the types and uses of various personal lighting devices.
    b. the required equipment, and location of external navigation lighting of the gyroplane.
    c. the meaning of various airport and navigation lights, the method of determining their status, and the procedure for airborne activation of runway lights.
2. Locates and identifies switches, spare fuses, and circuit breakers pertinent to night operations.

II. AREA OF OPERATION: PREFLIGHT PROCEDURES

A. TASK: PREFLIGHT INSPECTION

REFERENCE(S): FAA-H-8083-21; Gyroplane Flight Manual.

Objective. To determine that the applicant:

1. Exhibits knowledge of the elements related to a preflight inspection including which items must be inspected, the reasons for checking each item, and how to detect possible defects.
2. Inspects the gyroplane with reference to an appropriate checklist.
3. Verifies that the gyroplane is in condition for safe flight.

B. TASK: COCKPIT MANAGEMENT

REFERENCE(S): 14 CFR part 91; AC 91-32; FAA-H-8083-25; Gyroplane Flight Manual.

Objective. To determine that the applicant:

1. Exhibits knowledge of the elements related to efficient cockpit management procedures.
2. Ensures all loose items in the aircraft are secured.
3. Organizes and arranges material and equipment in an efficient manner so they are readily available.
4. Briefs the occupants on the use of safety belts, shoulder harnesses, doors, propeller and rotor blade avoidance, and emergency procedures.

C. TASK: ENGINE STARTING

REFERENCE(S): AC 91-13, AC 91-42, AC 91-55; FAA-H-808321, FAA-H-8083-25; Gyroplane Flight Manual.

Objective. To determine that the applicant:

1. Exhibits knowledge of the elements related to correct engine starting procedures. Including the use of an external power source, starting under various atmospheric conditions, awareness of other persons and property during start, and the effects of using incorrect starting procedures.

1. Positions the gyroplane properly considering structures, surface conditions, other aircraft, and the safety of nearby persons and property.
2. Utilizes the appropriate checklist for starting procedure.

D. TASK: TAXIING

REFERENCE(S): FAA-H-8083-21, FAA-H-8083-25; AIM; Gyroplane Flight Manual.

Objective. To determine that the applicant:

1. Exhibits knowledge of the elements related to recommended taxi procedures, including rotor blade management and the effect of wind during taxiing.
2. Performs a brake check immediately after the gyroplane begins moving.
3. Properly positions rotor blades while taxiing.
4. Controls direction and speed without excessive use of brakes.
5. Complies with airport markings, signals, ATC clearances, and instructions.
6. Avoids other aircraft and hazards.
7. Properly positions the gyroplane for runup considering other aircraft, surface conditions, and if applicable, existing wind conditions.

E. TASK: BEFORE TAKEOFF CHECK

REFERENCE(S): FAA-H-8083-21, FAA-H-8083-25; Gyroplane Flight Manual.

Objective. To determine that the applicant:

1. Exhibits knowledge of the elements related to the before takeoff check. Including, the reasons for checking the items and how to detect malfunctions.
2. Positions the gyroplane properly considering other aircraft, surface conditions, and wind conditions.
3. Divides attention inside and outside the aircraft.
4. Accomplishes the before takeoff check and ensures that the gyroplane is in safe operating condition.
5. Reviews takeoff performance airspeeds and expected takeoff distance.
6. Describes takeoff emergency procedures to include low speed/high speed blade flap situations.
7. Avoids runway incursions and/or ensures no conflict with traffic prior to taxiing into takeoff position.
8. Utilizes proper rotor spin-up procedure.

III. AREA OF OPERATION: AIRPORT OPERATIONS

A. TASK: RADIO COMMUNICATIONS AND ATC LIGHT SIGNALS

REFERENCE(S): 14 CFR part 91; FAA-H-8083-25; AIM.

Objective. To determine that the applicant:

1. Exhibits knowledge of the elements related to radio communications and ATC light signals.
2. Selects appropriate frequencies.
3. Transmits using recommended phraseology.
4. Acknowledges radio communications and complies with instructions.

B. TASK: TRAFFIC PATTERNS

REFERENCE(S): 14 CFR part 91; FAA-H-8083-25; AIM; Gyroplane Flight Manual.

Objective. To determine that the applicant:

1. Exhibits knowledge of the elements related to traffic patterns. Including, procedures at airports with and without operating control towers, prevention of runway incursions, collision avoidance, wake turbulence avoidance, and wind shear.
2. Complies with proper traffic pattern procedures.
3. Maintains proper spacing from other traffic.
4. corrects for wind drift to maintain proper ground tract.
5. Maintains orientation with the runway/landing area in use.
6. Maintains traffic pattern altitude, ±100 feet and appropriate airspeed, ±5 knots.

C. TASK: AIRPORT MARKINGS AND LIGHTING

REFERENCE(S): FAA-H-8083-25; AIM.

Objective. To determine that the applicant:

1. Exhibits knowledge of the elements related to airport runway and taxiway operations with emphasis on runway incursion avoidance.
2. Properly identifies and interprets airport runway and taxiway signs, markings, and lighting.

IV. AREA OF OPERATION: TAKEOFFS, LANDINGS, AND GO AROUNDS

A. TASK: NORMAL AND CROSSWIND TAKEOFF AND CLIMB

REFERENCE(S): FAA-H-8083-21; Gyroplane Flight Manual.

NOTE: If a calm wind weather condition exists, the applicant's knowledge of the crosswind elements must be evaluated through oral testing; otherwise a crosswind takeoff and climb must be demonstrated.

Objective. To determine that the applicant:

1. Exhibits knowledge of the elements related to normal and crosswind takeoff, climb operations, and rejected takeoff procedures.
2. Prerotates rotor blades to appropriate RPM.
3. Clears the area, taxies into the takeoff position, and aligns the gyroplane with takeoff path.
4. Advances the throttle as required.
5. Maintains proper directional control during acceleration on the surface.
6. Attains the proper lift-off attitude and airspeed.
7. Accelerates to appropriate climb airspeed, ±5 knots.
8. Maintains takeoff power to a safe maneuvering altitude, and then sets climb power.
9. Maintains directional control and proper wind-drift correction throughout the takeoff and climb.
10. Remains aware of the possibility of wind shear and/or wake turbulence.
11. Completes the prescribed checklist, if applicable.

B. TASK: NORMAL AND CROSSWIND APPROACH AND LANDING

REFERENCE(S): FAA-H-8083-21; Gyroplane Flight Manual.

NOTE: If a calm wind weather condition exists, the applicant's knowledge of the crosswind elements must be evaluated through oral testing; otherwise a crosswind approach and landing must be demonstrated.

Objective. To determine that the applicant:

1. Exhibits knowledge of the elements related to normal and crosswind approach and landing.
2. Adequately surveys the intended landing area.

3. Considers the wind conditions, landing surface, and obstructions. Selects a suitable touchdown point.
4. Establishes and maintains a stabilized approach at the recommended airspeed, with gust correction factor applied, ±5 knots.
5. Maintains proper ground track with crosswind correction, if necessary.
6. Remains aware of the possibility of wind shear and/or wake turbulence.
7. Makes smooth, timely, and correct control application during the flare and touchdown.
8. Touches down smoothly, beyond and within 50 feet of a specified point with no appreciable intended landing path.
9. Completes the prescribed checklist.

C. TASK: SOFT-FIELD TAKEOFF AND CLIMB

REFERENCE(S): FAA-H-8083-21; Gyroplane Flight Manual.

Objective. To determine that the applicant:

1. Exhibits knowledge of the elements related to a soft-field takeoff and climb.
2. Determines and utilizes best takeoff procedure based on the capabilities of this gyroplane and current conditions.
3. Positions the flight controls for existing wind conditions and to maximize lift as quickly as possible.
4. Prerotates rotor blades to appropriate RPM.
5. Clears the area, taxies onto the takeoff surface at a speed consistent with safety, without stopping, while advancing the throttle smoothly to takeoff power.
6. Maintains proper directional control.
7. Lifts off and remains in ground effect while accelerating to recommended climb airspeed.
8. Maintains recommended climb airspeed, ±5 knots.
9. Maintains takeoff power to a safe maneuvering altitude, then sets climb power.
10. Maintains proper ground track with crosswind correction, if necessary.
11. Remains aware of the possibility of wind shear and/or wake turbulence.
12. Completes the prescribed checklist.

D. TASK: SOFT-FIELD APPROACH AND LANDING

REFERENCE(S): FAA-H-8083-21; Gyroplane Flight Manual.

Objective. To determine that the applicant:

1. Exhibits knowledge of the elements related to soft-field approach and landing.
2. Considers the wind conditions, landing surface, and obstacles, and selects the most suitable touchdown area.
3. Establishes and maintains a stabilized approach at the recommended airspeed, with gust correction factor applied, ±5 knots.
4. Maintains proper ground track with crosswind correction, if necessary.
5. Remains aware of the possibility of wind shear and/or wake turbulence.
6. Makes smooth, timely, and correct control application during the flare and touchdown.
7. Touches down smoothly, at a minimum descent rate and airspeed with no appreciable drift, and with the longitudinal axis aligned with the intended landing path.
8. Completes the appropriate checklist.

E. TASK: SHORT-FIELD TAKEOFF AND CLIMB

REFERENCE(S): FAA-H-8083-21; Gyroplane Flight Manual.

Objective. To determine that the applicant:

1. Exhibits knowledge of the elements related to short-field takeoff and maximum performance climb.
2. Properly positions controls.
3. Prerotates rotor blades to appropriate RPM.
4. Clears the area, taxies into the takeoff position, and aligns the gyroplane for maximum utilization of available takeoff area.
5. Advances the throttle as required.
6. Climbs at manufacturer' s recommended airspeed, or in its absence at V, ± 5 knots until the obstacle is cleared, or until the gyroplane is at least 50 feet above the surface.
7. After clearing the obstacle, accelerates to appropriate airspeed, ±5 knots.
8. Maintains takeoff power to a safe maneuvering altitude, then sets climb power.
9. Maintains directional control and proper wind-drift correction throughout the takeoff and climb.

10. Remains aware of the possibility of wind shear and/or wake turbulence.
11. Completes the prescribed checklist, if applicable.

F. TASK: SHORT-FIELD APPROACH AND LANDING

REFERENCE(S): FAA-H-8083-21; Gyroplane Flight Manual.

Objective. To determine that the applicant:

1. Exhibits knowledge of the elements related to a short-field approach and landing.
2. Considers the wind conditions, landing surface, and obstacles.
3. Selects a suitable touchdown point.
4. Establishes and maintains a stabilized approach at the recommended airspeed, with gust correction factor applied, ±5 knots.
5. Maintains proper ground track with crosswind correction, if necessary.
6. Remains aware of the possibility of wind shear and/or wake turbulence.
7. Makes smooth, timely, and correct control application during the flare and touchdown.
8. Touches down smoothly, with little or no float beyond and within 50 feet of a specified point with no appreciable drift, and with the longitudinal axis aligned with the intended landing path.
9. Applies brakes, as necessary, to stop in the shortest distance consistent with safety.
10. Completes the prescribed checklist, if applicable.

G. TASK: GO-AROUND

REFERENCE(S): FAA-H-8083-21; Gyroplane Flight Manual.

Objective. To determine that the applicant:

1. Exhibits knowledge of the elements related to a go-around and when it is necessary.
2. Makes a timely decision to discontinue the approach to landing.
3. Applies appropriate power and establishes a climb at the appropriate airspeed, ±5 knots.
4. Maintains takeoff power to a safe maneuvering altitude, then sets climb power.
5. Maintains proper ground track with crosswind correction, if necessary.
6. Completes the prescribed checklist, if applicable.

V. AREA OF OPERATION: PERFORMANCE MANEUVER

A. TASK: STEEP TURNS

REFERENCE(S): FAA-H-8083-21; Gyroplane Flight Manual.

Objective. To determine that the applicant:

1. Exhibits knowledge of the elements related to steep turns.
2. Selects a safe altitude.
3. Establishes the manufacturers recommended airspeed or if one is not stated, a safe airspeed not to exceed VA.
4. Smoothly enters a coordinated steep 360° turn with a 40° bank, ±5°, immediately followed by at least a 360° turn in the opposite direction.
5. Divides attention between gyroplane control and orientation.
6. Maintains the entry altitude, ± 100 feet, airspeed, ± 10 knots, bank, ± 5° and rolls out on the entry heading ± 10° .

VI. AREA OF OPERATION: NAVIGATION

A. TASK: PILOTAGE AND DEAD RECKONING

REFERENCE(S): FAA-H-8083-25; AC 61-84.

Objective. To determine that the applicant:

1. Exhibits knowledge of the elements related to pilotage and dead reckoning.
2. Correctly flies to at least the first planned checkpoint to demonstrate accuracy in computations.
3. Identifies landmarks by relating the surface features to chart symbols.
4. Navigates by means of precomputed headings, groundspeed, and elapsed time.
5. Verifies the gyroplane's position within 1 nautical mile of flight planned route at all times.
6. Arrives at the en route checkpoints within 3 minutes of the initial or revised ETA and provides a destination estimate.
7. Maintains the appropriate altitude, ±100 feet and established heading, ±10°.

B. TASK: RADIO NAVIGATION AND RADAR SERVICES

REFERENCE(S): FAA-H-8083-25; AC 61-84; Navigation Equipment Operation Manuals.

NOTE: If the gyroplane is not equipped with radio navigation aids, competency will be evaluated through oral testing.

Objective. To determine that the applicant:

1. Exhibits knowledge of the elements related to radio navigation and ATC radar services.
2. Demonstrates the ability to use an airborne electronic navigation system.
3. Locates the gyroplane's position using the navigation system.
4. Intercepts and tracks a given course, radial, or bearing as appropriate.
5. Recognizes and describes the indication of station or way point passage as appropriate.
6. Recognizes signal loss and takes appropriate action.
7. Uses proper communication procedures when utilizing ATC radar services.
8. Maintains the appropriate altitude, ± 100 feet and headings ± 10°.

C. TASK: DIVERSION

REFERENCE(S): FAA-H-8083-21, FAA-H-8083-25; AC 61-84.

Objective. To determine that the applicant:

1. Exhibits knowledge of the elements related to procedures for diversion.
2. Selects an appropriate alternate airport and route.
3. Makes an accurate estimate of heading, groundspeed, arrival time, and fuel consumption to the alternate airport.
4. Maintains the appropriate altitude, ±100 feet and established heading, ±10°.

D. TASK: LOST PROCEDURES

REFERENCE(S): FAA-H-8083-25; AC 61-84; AIM.

Objective. To determine that the applicant:

1. Exhibits knowledge of the elements related to lost procedures.
2. Selects an appropriate course of action.
3. Maintains an appropriate heading, and climbs if necessary.
4. Identifies prominent landmarks.
5. Uses available navigation aids and/or contacts an appropriate facility for assistance, if gyroplane is radio equipped.
6. Plans a precautionary landing if deteriorating weather and/or fuel exhaustion is impending.

VII. AREA OF OPERATION: FLIGHT AT SLOW AIRSPEEDS

A. TASK: STRAIGHT-AND-LEVEL, TURNS, CLIMBS, AND DESCENTS AT SLOW AIRSPEEDS

REFERENCE(S): FAA-H-8083-21; Gyroplane Flight Manual.

Objective. To determine that the applicant:

1. Exhibits knowledge of the elements related to flight characteristics and controllability associated with maneuvering during slow flight.
2. Selects a safe altitude.
3. Establishes and maintains a specified airspeed, +5,-0, in straight-and-level flight, turns, climbs, and descents as directed.
4. Maintains the specified altitude, ±50 feet.
5. Maintains the specified heading during straight flight, ±5°.
6. Maintains specified bank angle, ±5°, during turning flight.
7. Rolls out on specified headings, ±5°.
8. Divides attention between gyroplane control and orientation.

B. TASK: HIGH RATE OF DESCENT AND RECOVERY

REFERENCE(S): FAA-H-8083-21; Gyroplane Flight Manual.

Objective. To determine that the applicant:

1. Exhibits knowledge of the elements related to aerodynamic factors associated with a high rate of descent and recovery and how this relates to actual approach and landing situations.
2. Selects an entry altitude that allows the task to be completed no lower than 500 feet AGL.
3. Establishes an airspeed that will induce a high rate of descent in high or low power settings.
4. Recognizes the onset of a high rate of descent.
5. Promptly recovers with or without power as directed.
6. Maintains the specified heading, ±10°.
7. Resumes normal cruising flight.

VIII. AREA OF OPERATION: EMERGENCY OPERATIONS

NOTE: TASK B may be tested orally at the discretion of the examiner, TASKs C through E are knowledge only items.

A. TASK: EMERGENCY APPROACH AND LANDING

REFERENCE(S): FAA-H-8083-21; Gyroplane Flight Manual.

Objective. To determine that the applicant:

1. Exhibits knowledge of the elements related to emergency approach and landing with a power failure.
2. Establishes and maintains the appropriate airspeed, ±5 knots.
3. Selects a suitable landing area, considering the possibility of an actual forced landing.
4. Plans and follows a flight pattern to the selected landing area, considering altitude, wind, terrain, obstacles, and other factors.
5. Attempts to determine the reason for the simulated malfunction, if time permits.
6. Completes the prescribed checklist, if applicable.

B. TASK: LIFT-OFF AT LOW AIRSPEED AND HIGH ANGLE OF ATTACK

REFERENCE(S): FAA-H-8083-21; Gyroplane Flight Manual.

Objective. To determine that the applicant:

1. Exhibits knowledge of the elements related to lift-off at low airspeed and high angle of attack, including combination of conditions, which are likely to lead to this situation.
2. Properly positions the controls.
3. Prerotates rotor blades to appropriate RPM, if applicable.
4. Clears the area, taxies into the takeoff position and aligns the gyroplane with the takeoff path.
5. Maintains proper directional control during acceleration on the surface.
6. Rotates for takeoff prior to normal lift-off airspeed with high angle of attack.
7. Detects the development of a low airspeed and high angle of attack, and initiates prompt corrective action.
8. Accelerates to appropriate climb airspeed, ±5 knots.

C. TASK: GROUND RESONANCE

REFERENCE(S): FAA-H-8083-25; Gyroplane Flight Manual.

Objective. To determine that the applicant:

1. Exhibits knowledge of the elements related to a fully articulated rotor system and the aerodynamics of ground resonance.
2. Understands the conditions that contribute to ground resonance.
3. Explains preventive flight techniques used during takeoffs and landings.

D. TASK: SYSTEMS AND EQUIPMENT MALFUNCTIONS

REFERENCE(S): Gyroplane Flight Manual.

Objective. To determine that the applicant:

1. Exhibits knowledge of the elements related to causes, indications, and pilot actions for various systems and equipment malfunctions.
2. Analyzes the situation and takes action, appropriate to the gyroplane used for the practical test, in at least four (4) of the following areas—
    a. engine/oil and fuel.
    b. hydraulic, if applicable.
    c. electrical.
    d. carburetor or induction icing.
    e. smoke and/or fire.
    f. flight control/trim.
    g. pitot static/vacuum and associated flight instruments, if applicable.
    h. rotor and/or propeller.
    i. any other emergency unique to the gyroplane flown.

E. TASK: EMERGENCY EQUIPMENT AND SURVIVAL GEAR

REFERENCE(S): FAA-H-8083-21; Gyroplane Flight Manual.

Objective. To determine that the applicant:

1. Exhibits knowledge of the elements related to emergency equipment appropriate to the gyroplane used for the practical test.
2. Identifies appropriate equipment that should be aboard the gyroplane.

IX. AREA OF OPERATION: POST-FLIGHT PROCEDURES

A. TASK: AFTER LANDING, PARKING AND SECURING

REFERENCE(S): FAA-H-8083-21, FAA-H-8083-25; AIM; Gyroplane Flight Manual.

Objective. To determine that the applicant:

1. Exhibits knowledge of the elements related to after-landing, parking, and securing procedures.
2. Maintains directional control after touchdown while decelerating to an appropriate speed.
3. Observes runway hold lines and other surface control markings and lighting.
4. Parks in an appropriate area, considering the safety of nearby persons and property.
5. Follows the appropriate procedure for engine shutdown.
6. Completes the appropriate checklist.
7. conducts an appropriate post flight inspection and secures the aircraft.

APPENDIX 1
TASK VS. SIMULATION DEVICE CREDIT

FLIGHT TASK Areas of Operation :	1	2	3	4	5	6	7	A	B	C	D
II. Preflight Procedures											
A. Preflight Inspection (Cockpit Only)											
B. Cockpit Management											
C. Engine Starting and Rotor Engagement (If applicable)											
D. Before Takeoff Check											
III. Airport and Heliport Operations											
A. Radio Communications and ATC Light Signals											
B. Traffic Patterns											
C. Airport/Heliport Runway, Heliport, and Taxiway Signs, Markings, and Lighting											
IV. Hovering Maneuvers											
A. Vertical Takeoff and Landing											
B. Slope Operations											
C. Surface Taxi											
D. Hover Taxi											
E. Air Taxi											
V. Takeoffs, Landings, and Go-Arounds											
A. Normal and Crosswind Takeoff and Climb											
B. Normal and Crosswind Approach											
C. Maximum Performance Takeoff and Climb											
D. Steep Approach											
E. Rolling Takeoff											
F. Shallow Approach and Running /Roll-On											
G. Go-Around											
VI. Performance Maneuvers											
A. Rapid Deceleration											
B. Straight In Autorotation											
C. 180° Autorotations											
D. Approach and Landing with Simulated Powerplant Failure – Multiengine Helicopter											
VII. Navigation											
A. Pilotage and Dead Reckoning										X	X
B. Radio Navigation and Radar Services										X	X
C. Diversion									X	X	
D. Lost Procedures										X	X
VIII. Emergency Operations											
A. Power Failure at a Hover										X	X
B. Power Failure at Altitude										X	X
C. Systems and Equipment Malfunctions										X	X
D. Settling-With-Power										X	X
E. Low Rotor RPM Recovery										X	X
F. Dynamic Roll Over											
G. Ground Resonance											
H. Low G Conditions											
I. Emergency Equipment and Survival Gear											
IX. Special Operations											
A. Confined Area Operations											
B. Pinnacle/Platform Operations											
X. Postflight Procedures											
A. After Landing and Securing											

X1 Creditable only if accomplished in conjunction with a running takeoff or running landing, as appropriate.

NOTE: 1. The helicopter may be used for all TASKs.

2. Level C simulators may be used as indicated only if the applicant meets established pre-requisite experience requirements.

3. Level A helicopter simulator standards have not been defined.

4. Helicopter flight training devices have not been defined.

Made in United States
Troutdale, OR
11/03/2026

56844443R00049